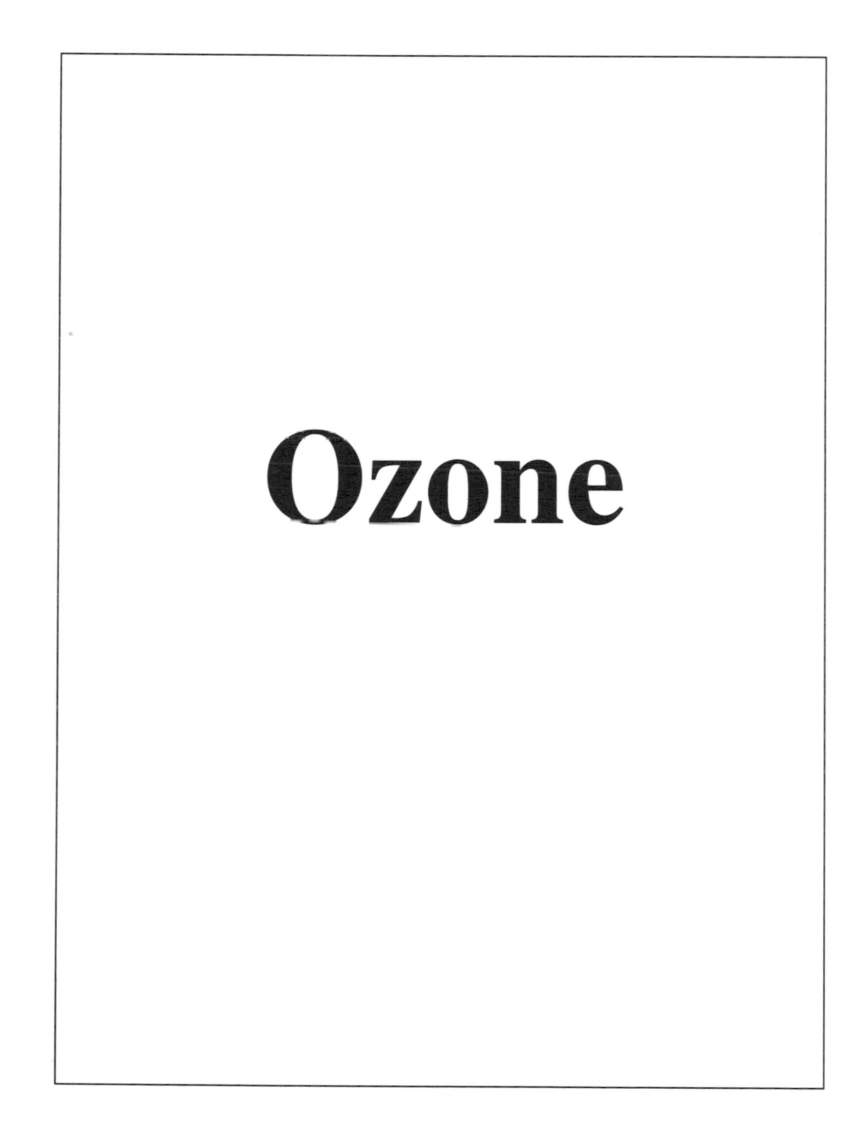
Ozone

Look for these and other books in the Lucent Overview series:

Abortion
Acid Rain
AIDS
Animal Rights
The Beginning of Writing
Cancer
Dealing with Death
Drugs and Sports
Drug Trafficking
Eating Disorders
Endangered Species
Energy Alternatives
Extraterrestrial Life
Garbage
Gun Control
Hazardous Waste
The Holocaust
Homeless Children
Ocean Pollution
Oil Spills
The Olympic Games
Ozone
Population
Rainforests
Smoking
Special Effects in the Movies
Teen Alcoholism
The UFO Challenge
Vietnam

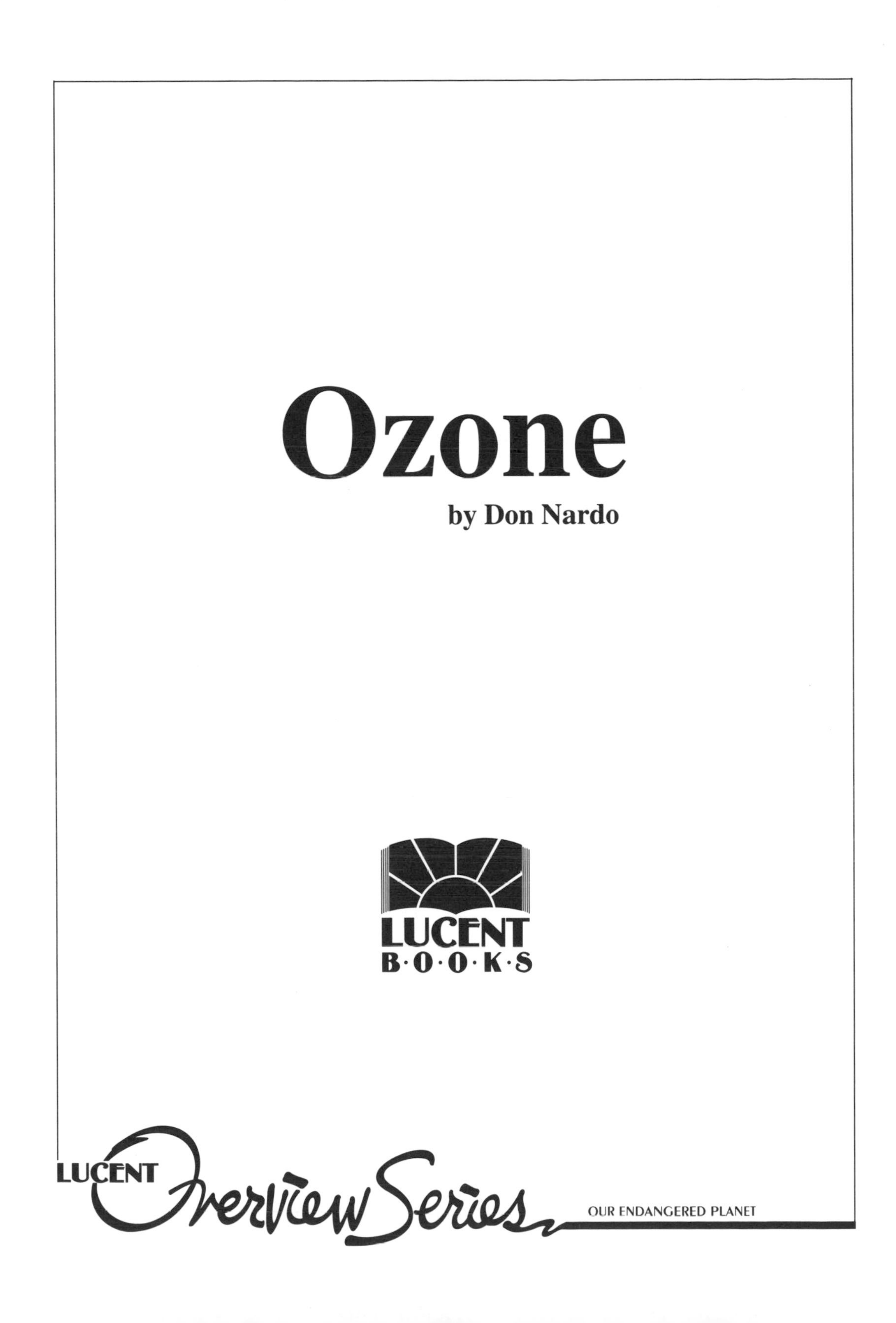

Ozone

by Don Nardo

LUCENT BOOKS

LUCENT Overview Series

OUR ENDANGERED PLANET

LUCENT Overview Series OUR ENDANGERED PLANET

Library of Congress Cataloging-in-Publication Data

Nardo, Don, 1947-
Ozone / by Don Nardo.
p. cm. — (Lucent overview series. Our endangered planet)
Includes bibliographical references and index.
Summary: Discusses the nature and necessity of ozone in relation to the global environment and air pollution, and examines the greenhouse effect and ways to prevent it.
ISBN 1-56006-101-4
1. Atmospheric ozone—Environmental aspects—Juvenile literature. 2. Greenhouse effect, Atmospheric—Juvenile literature. 3. Air—Pollution—Juvenile literature. [1. Ozone. 2. Ozone layer. 3. Greenhouse effect, Atmospheric. 4. Air—Pollution.] I. Title.
TD1030.5.G36 1991
363.73'92—dc20 91-6275

P.O. Box 289011 San Diego, CA 92198-9011

Contents

Introduction

(opposite page) *The haze that usually hangs over Los Angeles, California was especially thick during the summer of 1988, when record temperatures increased the amount of ozone in the air. Too much ozone in the lower atmosphere is harmful to people, animals, and plants.*

THE SUMMER OF 1988 was unusually hot in North America. All across the United States, temperatures soared, breaking records and boosting sales of fans and air conditioners. A brownish yellow haze hung over Los Angeles, California, nearly every day. People in Los Angeles are used to this kind of pollution. They call it smog. But the smog during the summer of 1988 seemed worse than ever. Millions of people coughed and wheezed, and their eyes watered and stung. City officials issued smog alerts, warning people to stay indoors as much as possible. Many of the smog sufferers were not aware that their problems were due, in large part, to an invisible gas called ozone. Too much ozone in the air was causing the severe pollution.

While North America sizzled through the summer of 1988, the residents of Australia, located below the equator, experienced winter. Australian officials issued alerts, too. Like the residents of Los Angeles, people in the land Down Under were warned to stay indoors. But these warnings had nothing to do with smog. There, the problem was too many ultraviolet rays from the sun. The Australians knew that too much exposure to these ultraviolet rays could cause skin cancer and other illnesses. High ultraviolet levels had become com-

mon in Australia. Part of the problem was that in this case there was not enough ozone gas in the air.

Too much ozone in the air? Not enough in the air? This seems like a contradiction, but it is not. Ozone gas has a strange double nature and the power to significantly affect the environment and the lives of people everywhere.

(opposite page) In Australia, sunbathers like this one must protect themselves from intense ultraviolet radiation. The incidence of malignant melanoma, the most dangerous form of skin cancer, is four times higher in Australia than in the United States.

1

What Is Ozone?

IN 1839, THE GERMAN-BORN scientist Christian Friederich Schönbein was a professor of chemistry at the University of Basel in Switzerland. In addition to his teaching duties, he regularly performed laboratory experiments in hopes of finding out which substances make up the earth's atmosphere. To do this, Schonbein combined and separated various liquids and gases.

During one of these experiments, Schonbein passed an electric charge through a flask of water. Each time he passed the electric charge through the water, he noticed a peculiar odor. The scientist first thought that the smell must be electrically charged oxygen. But finally, he realized he had found a new substance. He named this new substance ozone, after the Greek word *ozein,* meaning "to smell."

The properties of ozone

Ozone is a gas and is a form of oxygen. Each molecule of ozone contains three atoms of oxygen. Molecules are tiny building blocks that form the structure of all solids, liquids, and gases. Each molecule is formed from still smaller building blocks called atoms, which are the basic components of all matter. Because each molecule of ozone is composed of three oxygen atoms, scientists refer to ozone as O_3.

(opposite page) The electrical smell that often lingers after a thunderstorm is the smell of ozone. During electrical storms, lightning cuts through the air, separating oxygen molecules. The oxygen atoms recombine to form ozone.

Rust forms on drums like these when ozone molecules break apart and recombine with iron atoms during the process of oxidation.

Ozone gas is colorless but has a strong odor. The "electrical" smell that often lingers after a summer thunderstorm is the smell of ozone. The ozone molecule is unstable; it has a tendency to break apart and join with other atoms. This process, called oxidation, can be destructive. Rust develops on iron, for example, when the oxygen atoms in ozone combine with iron atoms. Similarly, ozone can weaken materials like nylon and rubber. In large amounts, ozone can even kill living cells. This is why scientists must be able to measure and carefully monitor ozone in the air.

Measuring ozone

Schönbein himself invented a crude way of measuring levels of ozone concentration. First, he soaked a piece of paper with a chemical. As the chemical dried, the ozone in the air oxidized it. This turned the paper blue. The more ozone present, the darker blue the paper became. Schönbein placed a new piece of paper outside every day in order to measure the levels of ozone in the air. In time, scientists came to refer to this technique as the Schönbein paper method. Unfortunately, this method shows only whether ozone levels are light, moderate, or heavy, so it is not very precise.

A better method for measuring ozone levels was developed in 1876. Charles Soret, a French scientist who had been studying weather and ozone, decided that the Schönbein papers were not always accurate. So Soret devised his own approach. He measured ozone by observing the way it reacted with chemicals dissolved in water. Using this method, he was able to more accurately measure ozone.

Soret measured the air nearly every day for thirty-four years. Later, scientists used Soret's work to compare ozone levels of the past and present.

Modern scientists have an even more accurate method for measuring ozone. This method is called gas chromatography. This process works by filtering and burning gases containing ozone. The device used in this process is so precise that it can detect a single molecule of ozone among one million other molecules.

How ozone forms

There are two general types of ozone: low-level and upper-level ozone. Low-level ozone exists near the earth's surface in the lowest portion of the atmosphere. This is why scientists refer to it as low-level ozone. This lowest atmospheric layer, occupying the space between the earth's surface and an altitude of about six to nine miles, is called the troposphere. Scientists also refer to low-level ozone as tropospheric ozone.

Tropospheric ozone, which can be harmful to living things, is created in many ways. Usually, ozone is produced when sunlight reacts with chemicals released into the air. Sometimes, human activities are responsible for the creation of ozone. Exhausts from gasoline-burning cars and trucks, for example, contain various pollutants, including hydrocarbons and nitric oxide. These substances react with each other and also with sunlight, creating ozone in the process. Emissions from coal-, gas-, and oil-burning utility plants and vapors from paint strippers, dry-cleaning methods, and charcoal lighter fluids also produce ozone when they mix with sunlight.

Fire, whether it occurs naturally or is set by human beings, is also a source of low-level ozone. The smoke from fire contains chemicals that produce ozone when they mix with sunlight. During the sugarcane harvest in Brazil, for instance, farmers often burn off the excess vegetation that grows around the cane stalks. In some years, more than twenty thousand cane fires are set each week during the harvest season. Because Brazil is near the equator, it receives the sun's rays more directly than do areas farther north or south. This combination of pollutants released by the fires and plenty of direct sunlight produces huge amounts of ozone pollution.

Backyard barbecues produce low-level ozone when the chemicals in the smoke and lighter fluid react with sunlight.

Natural ozone

Other sources of tropospheric ozone are completely natural. During electrical storms, bolts of lightning rip through the troposphere, separating the atoms of some oxygen molecules. Some of these atoms combine with other oxygen molecules to create ozone. There are other ways nature makes low-level ozone. Ozone forms when sunlight combines with methane, a gas given off by decaying plant and animal tissue and by the gaseous wastes of grazing animals like cows and sheep.

Another natural source of tropospheric ozone was not fully understood by scientists until the late 1980s. In 1987, William Chameides, Ronald Lindsay, and Jennifer Richardson, all of the Geor-

Sunlight reacts with the hydrocarbons and nitric oxide in car exhaust to form ozone.

gia Institute of Technology, studied ozone levels around Atlanta, Georgia. They found that the amount of ozone in the air was far too much to be attributed to cars and factories alone. The researchers finally concluded that trees in the Atlanta area were creating ozone. The researchers found that trees released hydrocarbons during photosynthesis. This is the process by which green plants use sunlight to create energy. When the hydrocarbons reacted with sunlight and pollutants, they created ozone.

Trees generate large amounts of hydrocarbons. In fact, Chameides and the others found that the amount of hydrocarbons released by trees in the Atlanta region was about the same as the amount emitted by cars and industrial facilities. This does not mean that trees pollute. Atmospheric scientist Jack Fishman of the National Aeronautics and Space Administration's Langley Research Center, explains: "Actually, trees don't pollute. . . . In a natural forest environment, the concentration of [pollutants from engine exhausts] is usually very low, so the hydrocarbon given off by trees is harmlessly transformed into substances other than ozone." Only when the hydrocarbons released by the trees combine with other pollutants given off by the fuel-burning activities of humans is too much ozone produced.

Upper-level ozone

The second general type of ozone exists in the upper portion of the earth's atmosphere and is called upper-level ozone. This layer of the atmosphere, extending from about twelve to twenty-four miles above the planet's surface, is known as the stratosphere. Therefore, scientists also refer to upper-level ozone as stratospheric ozone. This ozone layer blocks the most dangerous of the sun's rays and prevents them from harming living things on earth. This role as protector of life below began billions of years ago when there was

Emissions from industrial plants also produce ozone when they react with sunlight. Car exhausts and industrial pollutants account for half of the hydrocarbon emissions that contribute to the formation of ozone.

little oxygen in the air.

When ocean plants first appeared on earth, most had to stay below the water's surface. This was to avoid ultraviolet light, a kind of solar ray that damages living tissue. Ultraviolet light flooded the primitive earth, making the development of life a slow, painstaking process.

As the oxygen given off by the early plants slowly built up in the atmosphere, some of it floated high into the stratosphere. There, the oxygen molecules encountered sunlight that was stronger than what reached the surface. This combination of sunlight and oxygen molecules produced ozone, a gas that could readily absorb ultraviolet light. With protection provided by the ozone layer, plants thrived in the ocean and on its surface. The amount of oxygen in the air increased and formed more ozone. Eventually, the stratospheric ozone layer grew thick enough to protect more advanced forms of life from ultraviolet radiation. Life evolved onto land and into the millions of varieties of plant and animal species known today, including human beings.

Two faces of ozone

The nature of ozone appears to go against common sense. On the one hand, there is low-level, tropospheric ozone, which is classified as a pollutant. Then, there is the upper-level, stratospheric ozone that is essential to the existence of life.

The amount of both kinds of ozone is changing. Low-level ozone is increasing, while upper-level ozone is decreasing. Scientists disagree on what these changes mean. But, it appears that there is too much of one kind of ozone and too little of the other. Researchers are trying to determine how these levels of ozone will affect the world environment and people's lives.

2

Too Much Ozone

TOO MUCH LOW-LEVEL ozone can be harmful to people, animals, and plants. Highly concentrated ozone is so lethal that it is used to kill germs on medical equipment. In lower concentrations, ozone can cause breathing difficulties for humans and other problems for plants. Scientists have known about the potential dangers of ozone for decades. But not until the 1970s and 1980s did ozone concentrations near the earth's surface reach high levels often enough to cause much concern.

(opposite page) This woman wears a gas mask to protest destructive environmental practices, on Earth Day, 1970.

The long, hot summer

Average low-level ozone concentrations rose steadily in the 1980s. The worst large-scale incident of tropospheric ozone pollution occurred during the summer of 1988. A huge mass of warm air hung over the eastern half of the United States for most of the summer. There was so little air movement in the lower atmosphere that ozone levels built up rapidly. The air was so bad in Detroit, Michigan, for example, that ozone levels exceeded the amount considered safe by the U.S. Environmental Protection Agency (EPA) sixteen times. Ozone levels in Philadelphia, Pennsylvania, exceeded EPA limits forty-two times that summer. And cities were not the only areas that suffered.

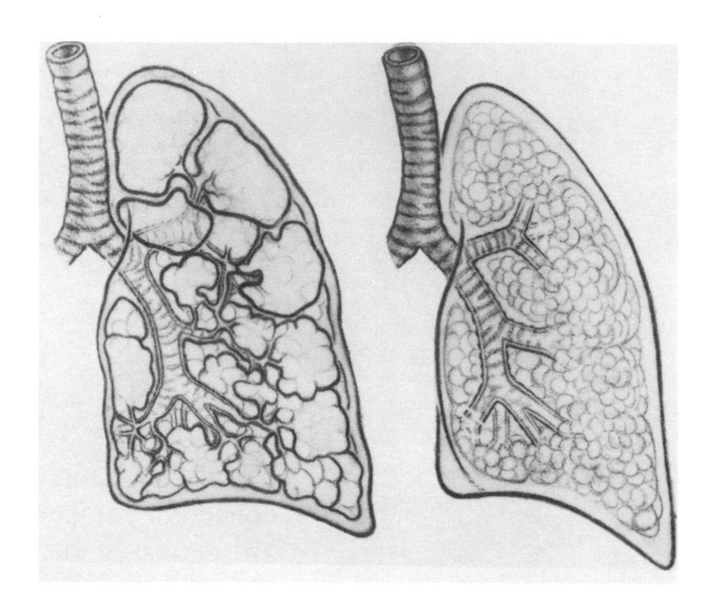

Some scientists believe that continued exposure to high concentrations of ozone can lead to emphysema. The lung on the left is infected with emphysema, a disease in which a person cannot get enough oxygen from normal breathing.

Ozone levels broke records in the Shenandoah Valley of Virginia. In Maine's Acadia National Park, ozone concentrations were the highest ever, and park rangers warned people not to exert themselves.

Because of the rise in ozone pollution in the 1980s, scientists began paying more attention to ozone. Many studies were done. They revealed some new information about the harmful effects of low-level ozone.

Ozone and lung damage

Some studies showed that ozone damages the lungs of animals and people. Inside the lungs are thousands of tiny air sacs through which oxygen and carbon dioxide pass when people breathe. Scientists have found that the thin walls of these air sacs suffer the most damage from ozone. Scientists are not completely sure how this occurs. One theory suggests ozone molecules combine with air sac molecules, creating scar tissue. The scar tissue eventually causes breathing problems. The other theory contends that the ozone reacts with fatty tissue in the air sacs, hardening the tissue and causing similar breathing problems. In either case, continued exposure to high concentrations of ozone can lead to emphysema, a lung disease in which a person cannot get enough oxygen from normal breathing.

Determining safety standards

If too much ozone can be dangerous to a person's health, exactly how much ozone is too much? In 1971, the EPA set a national safety standard that allowed eight molecules of ozone for every billion air molecules. This meant that exposure to less than this amount was considered safe. In 1977, the EPA upped the standard to twelve.

Some later studies suggested that this new standard might be too high. For instance, scientists exposed baboons, rabbits, and dogs to various ozone levels. Exposures of one-third the EPA standard caused at least minor inflammation, red-

This man wears a surgical mask to protect himself from the severe pollution experienced daily in Mexico City.

ness, and swelling of lung tissues. A 1988 study by researcher Richard Mann and others tested human beings and found similar results. And an EPA lab in South Carolina recently conducted a study that took into account the factor of exercise. This is important because the more vigorously a person exercises, the more air, and therefore the more ozone, he or she inhales. The researchers exposed people to various levels of ozone. Each time the subjects were exposed, they exercised for a set length of time on a treadmill or stationary bicycle. The results of this experiment showed that when people exercised vigorously lung damage occurred at ozone levels lower than the EPA standard.

Some researchers disagree with these findings. Economist Lawrence White conducted a survey of studies on ozone and health and concluded that ozone levels nearly three times those allowed by the EPA would not be unhealthy unless a person exercised very strenuously. And even then, he said, the negative effects might be due to the ex-

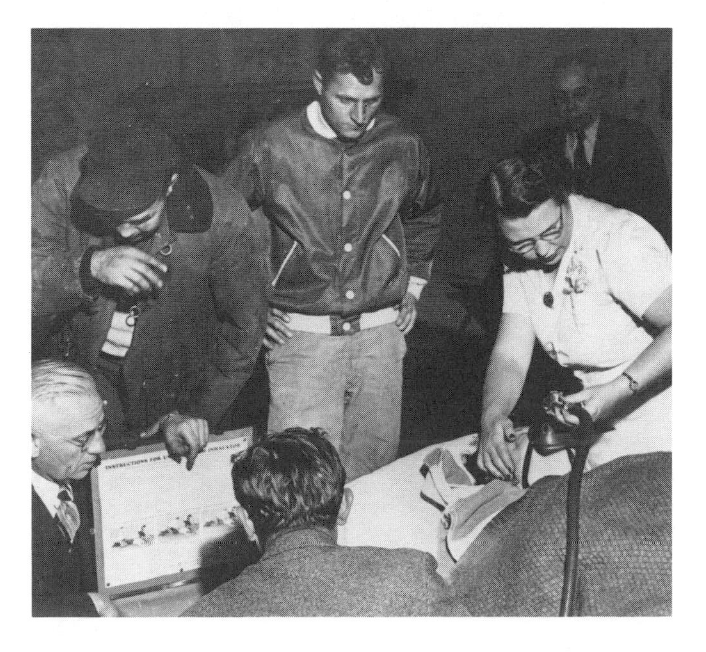

In Donora, Pennsylvania, a nurse administers oxygen during the town's industrial-smog disaster in 1948. Seventeen people died after breathing the fumes from local steel mills.

ercise itself. White said, "Thus far, ozone exposure has not been demonstrated to have long-term debilitating [harmful] consequences in humans."

Unanswered questions

By contrast, studies conducted in the early 1980s in Pennsylvania and New Jersey found harmful health effects at much lower ozone levels. The researchers examined active children at summer camps to see if ozone affected their lungs. They found a definite increase in breathing problems among some of the children, even at levels near the EPA standard. But other children seemed relatively unaffected by the ozone. This suggested to the researchers that ozone affects some people more than it does others. This idea gained support when the researchers found that the girls studied were more sensitive to ozone than the boys. The reason for this is still unclear.

Most ozone researchers agree that many questions about safe ozone levels remain unanswered and that more extensive studies are needed. However, some people suggest that ozone should not be viewed as a major problem. Melinda Warren and Kenneth Chilton of the Center for the Study of American Business at Washington University in St. Louis, Missouri, sum up this view:

> Ozone's health effects are relatively mild and temporary. . . . The ability of people to adapt is greater than generally recognized. . . . There is no evidence that people with preexisting lung disease are affected more severely than healthy persons by high concentrations of ozone. . . . The point is not that ozone pollution should be ignored but rather that its health consequences need to be kept in perspective.

Those who agree with this statement do not believe that unlimited concentrations of ozone are safe. Rather, they argue that the safety limit—the number of parts per billion that humans can tolerate—may be much higher than most scientists believe.

The one thing all the researchers agree on is that the higher tropospheric ozone levels rise, the more they threaten human health.

Ozone and the plant kingdom

Low-level ozone is also harmful to plants and crops. Scientists have studied the effects of ozone on plants for several decades. Most of the early studies were laboratory experiments that concentrated on the damage ozone "might" do if levels became high enough. Since no one at that time had ever knowingly experienced dangerous ozone levels, there seemed little cause for concern.

Concentrated ozone prevents the leaves of trees from absorbing the nutrients they need from the air. In severe cases, trees will die.

Eventually, however, tropospheric ozone levels *did* become high enough to cause problems. In the 1970s and 1980s, ozone levels in many areas of the world rose sharply, and people began to see widespread damage to forests and crops. For instance, several forests in the state of Bavaria in Germany showed signs of "early autumn syndrome." This is a condition in which the trees grow sick and lose their leaves by late August instead of during the autumn season. Outbreaks of the syndrome also began to increase in the United States and other countries. As a result, scientists started studying the problem.

Like animals and people, plants have an organ that "inhales" gases from the air and "exhales" waste materials. This organ is the leaf. A plant leaf is covered with millions of tiny openings through which the plant "breathes." When leaves are exposed to high concentrations of ozone, these openings close, making it impossible for the plant to take in the gases and nutrients it needs to live. If trees experience this condition long enough, early autumn syndrome sets in. In the

In this 1953 photo, a tourist wears a mask to protect herself against London's polluted air. In 1952, stalled air trapped smoke and sulfur dioxide from nearby industrial plants over the city. Over four thousand people died of respiratory illnesses.

most severe cases, the trees die.

Like trees and other forest plants, food crops can suffer from exposure to low-level ozone. The extent of the damage, however, and the amount of money lost each year from that damage are unclear. The EPA estimates that crop losses from ozone in the United States total $2.5 billion to $3 billion per year. The World Resources Institute, an environmental research organization, believes that annual losses in the United States are as high as $5 billion, while some university studies have concluded that the yearly damage amounts to about $1 billion. Many scientists say more research is needed.

Another question researchers are still unable to answer is why ozone hurts some crops and not

Dr. Joe Shilen (standing) and Dr. J.S. Sharrah conduct air-quality tests after seventeen people died during the Donora, Pennsylvania smog crisis in 1948.

others. Studies have shown that exposure to high levels of ozone increases the chances of some crops catching plant diseases. Potatoes and barley both become diseased more easily after contact with concentrated ozone. Yet the studies also show that tomatoes and cabbage actually get sick less after being exposed to ozone. This curious fact has led some researchers to suggest that the worst damage to the plants may not be from ozone alone. Instead, they say that much of the crop damage may be caused by a mixture of ozone and other pollutants.

Sulfur dioxide

One air pollutant that scientists have studied in combination with low-level ozone is sulfur dioxide, a by-product of burning coal. Even by itself, sulfur dioxide can be dangerous to life. The world learned this in 1952 when a mass of warm air stalled over London, England. Smoke and sulfur dioxide from nearby coal plants were trapped over the city by unmoving air. The result was the worst air pollution tragedy ever recorded. More than four thousand people died of respiratory illnesses.

In 1966, researchers H.A. Menser and H.E. Hoggestad exposed tobacco plants first to ozone alone, then to sulfur dioxide alone, each for a period of four hours. The plants did not appear to be hurt in either case. But, when the researchers exposed the plants to a mixture of the two pollutants for four hours, nearly half of the leaves shriveled and died. Later studies, including one conducted at EPA labs, showed that broccoli, tomatoes, apples, grapes, soybeans, and many other crops suffered similarly from the combination of ozone and sulfur dioxide.

It might seem that the threat to crops is serious only on farms located near cities. Cities, after all, are usually more polluted than the countryside because of the higher concentration of cars, trucks, and industrial plants in the cities. But scientists have learned that wind can carry ozone pollution over great distances. Rural sections of Virginia and Maine measured record ozone levels in the summer of 1988. According to researcher Jack Fishman:

> Such occurrences in nonindustrial states like Virginia and Maine underscore one of the most distressing problems of ozone. . . . The polluted air over those states, for the most part, didn't originate there. The ozone levels experienced in Maine originated to the south and southwest, and the ozone in Virginia may have originated in the [oil] refineries and industrialized areas of the Texas and Louisiana Gulf Coast. Some studies have shown that some of the pollution measured over southeastern Virginia originated in the New York-Boston [air] corridor.

As ozone and sulfur dioxide levels continue to rise in U.S. cities and industrial areas, the pollutants are a potential threat to crops throughout the country.

Smoke billows from this plant in Champaign, Illinois. The great number of cars and industrial plants causes most city pollution.

Smog

The word *smog*, a combination of the words *smoke* and *fog*, originally described the thick brown haze that hung over many industrial cities in the nineteenth century. This haze was composed of pollutants such as carbon dioxide, sulfur dioxide, and soot that poured from factory smokestacks. Some cities suffered severely from this industrial smog. The "killer smog" that struck London in 1952 is one example. A similar smog-induced disaster occurred in the steel-mill town of Donora, Pennsylvania, in 1948, killing seventeen people. And in the early twentieth century, Pittsburgh, Pennsylvania, earned the nickname "smoke city" because its inhabitants frequently suffered from industrial smog.

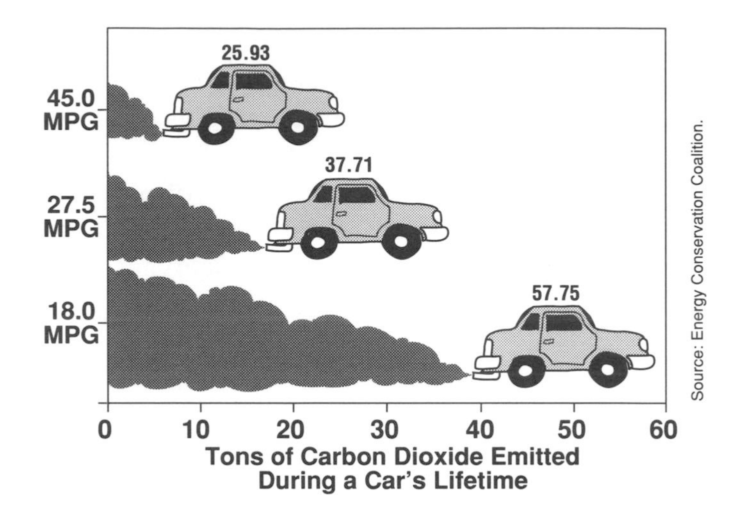

The amount of carbon dioxide emitted during a car's lifetime depends on fuel efficiency (miles per gallon). A car that is fuel efficient releases less carbon dioxide than one that is not.

By contrast, in modern cities, industrial smog mixes with polluting exhausts from cars and trucks and then with sunlight to create ozone smog.

The American city that has suffered the most from ozone smog is Los Angeles. In the late 1930s, the city began to expand, and by 1945, it was the fastest-growing city in the United States. In the 1940s, Los Angeles and the state of California built thousands of miles of highways in and around the city to keep up with the swiftly growing population. By the early 1960s, there were more cars and trucks in southern California than in any other area of similar size in the United States. Los Angeles then had one of the three factors necessary for creating severe ozone smog—plenty of car and truck exhaust.

Once the fastest-growing city in the nation, Los Angeles built thousands of miles of highways to keep up with its expanding population. This stretch of the Santa Monica freeway is considered the busiest in the world.

Unfortunately, the city already had the other two factors. Los Angeles rests in a flat basin surrounded by mountains, except on the side facing the ocean. This often allows warm air to stall and hang over the city, sometimes for days at a time. The other factor is sunlight. Southern California has a pleasant climate with hundreds of days of bright sunshine each year.

These three factors—large amounts of car exhaust, unique terrain, and lots of sunlight—combine to produce serious ozone smog in Los Angeles. By the mid-1960s, ozone levels in the city often measured more than three times the level now regarded as safe by the EPA. City officials frequently issued "smog alerts," advising children and elderly people to stay indoors or decrease their amount of outdoor activity. The officials warned all citizens not to exert themselves. A thick haze sometimes hid tall buildings and nearby mountains from view.

Los Angeles eventually reduced its low-level

Los Angeles, California, rests in a basin, allowing warm air to stall over the city, sometimes for days at a time.

ozone. Air pollution controls in the 1970s got rid of many of the pollutants released by cars and trucks. Although ozone alerts now happen less frequently in the city, ozone levels remain high, regularly violating the EPA standard.

Los Angeles is not alone in failing to meet the EPA's low-level ozone standard. During the summer of 1988, for instance, ninety-six cities and counties in the United States repeatedly went over the level considered safe by the EPA.

The problem is even worse in many foreign countries. Ozone smog hangs over dozens of cities from Athens to Tokyo. Even some rural areas are threatened by ozone smog. In Thailand, Brazil, and some countries in eastern Africa, the burning of crops to enrich the soil and the burning of forests to clear land regularly produce high

Dangerous levels of smog threaten many U.S. cities. Here, emissions from an oil refinery in Houston, Texas, pollute the air.

levels of ozone. The smog often becomes so thick that it reduces the brightness of the midday sun to a faded twilight glow.

This part of the rainforest in Ivory Coast, Africa was burned to make the land available for farming. The smoke from such fires reacts with sunlight to produce high amounts of ozone.

The ozone machine

Many ways of creating low-level ozone exist in the modern industrial world. According to the EPA, in 1987 alone, cars, factories, fires, waste dumps, and other pollution sources in the United States poured more than ninety million tons of ozone-producing substances into the air. In addition, more than eighteen million tons of sulfur dioxide entered the atmosphere above the United States, creating the potential for a dangerous mixing of that gas and ozone. Scientists estimate that trees produce at least several tens of millions of tons of hydrocarbons each year. Many of these hydrocarbons create ozone when they mix with car exhaust. The addition of huge amounts of sunlight to these substances makes the earth's lower atmosphere, in a sense, one big ozone-producing "machine."

The exact amount of tropospheric ozone produced by the ozone machine is still uncertain. Scientists are also unsure of just how much ozone is dangerous to animals and people. And there is still much disagreement about the size of crop losses due to ozone damage.

But scientists do agree on two important points. First, highly concentrated low-level ozone is harmful to living things and alters or destroys many of the substances with which it comes in contact. Second, the world's overall levels of tropospheric ozone are steadily rising. Unfortunately, it is impossible for people to move this excess ozone to higher atmospheric levels where the problem is not too much, but too little ozone.

3

Too Little Ozone

(opposite page) *Thousands gather on this Santa Monica, California, beach to enjoy 90 degree heat and the Fourth of July holiday. Scientists worry that as the ozone shield weakens and the intensity of ultraviolet radiation increases, cancer rates and deaths will also increase.*

WHILE TOO MUCH low-level ozone can be harmful, exactly the opposite is true for upper-level ozone. Because ozone high in the stratosphere absorbs incoming ultraviolet light from the sun, it keeps most of these destructive rays from reaching the earth's surface. Ultraviolet radiation is so lethal that without the stratospheric ozone layer, the existence of life on earth would be nearly impossible.

Until the 1970s and 1980s, most people—including scientists—took the ozone layer more or less for granted. They assumed it was one of the earth's natural features and that it would always remain the same. This attitude changed in 1984 when scientists found a huge hole in the ozone layer over Antarctica—the continent covering the planet's south pole. Since then, researchers have discovered that the stratospheric ozone shield is growing thinner and weaker. As this happens, more ultraviolet light reaches the earth's surface. This is why scientists and government officials around the world are increasingly concerned about the possible health effects of upper-level ozone depletion.

At the beach, when people lie under a warm summer sun, many leave with attractive golden tans. Others, especially those with fair skin, get painful sunburns. Both the tans and sunburns are caused by ultraviolet rays striking and damaging the cells on the surface of the skin.

Scientists and doctors have known for a long time that repeated exposure to ultraviolet radiation causes skin cancer. In general, light-skinned people have the highest risk, while dark-skinned people have a lower risk. This is because darker skins contain more of the pigment melanin, which blocks most ultraviolet light. About 90 percent of all skin cancers occur on the head and neck, mainly because these are the areas of the body that are most often left uncovered and exposed to sunlight. As might be expected, skin cancer rates are highest for people who frequently work or play outside.

About 400,000 to 600,000 cases of skin cancer occur in the United States each year. Many of these cases are caused by too much exposure to ultraviolet rays from the sun. Most skin cancer, if

Tans and sunburns are caused by ultraviolet rays striking and damaging cells on the surface of the skin.

Protective garb helps to shield workers who must work in the out-of-doors. About 90 percent of all skin cancers occur on the head and neck—the areas that are most often left uncovered.

caught early, is curable. But about six thousand people still die each year from the disease. As the intensity of ultraviolet radiation increases, cancer rates and deaths may also increase. Many scientists expect that as upper-level ozone decreases, the intensity of ultraviolet radiation at the earth's surface will increase. According to the EPA, for every 1 percent decrease in the ozone layer, ultraviolet rays reaching the earth's surface will increase 2 percent. This, in turn, could cause an 8 percent rise in skin cancer rates or another 80,000 to 100,000 cases of skin cancer in the United States.

Trouble Down Under

Such increases in skin cancer rates are not just a matter of scientific predictions and estimates. They are already happening. The most widely publicized examples are in Australia. That country is the closest heavily populated area to Antarctica, where upper-level ozone is disappearing. After the discovery of the Antarctic ozone hole, many scientists predicted that Australia would show the highest rates of skin cancer in the world. And this is exactly what happened.

Studies have shown that the incidence of malignant melanoma, the most dangerous form of skin cancer, is four times higher in Australia than in the United States. About one out of every four Australians suffering from malignant melanoma dies from the disease. Researchers caution that not all of these cases are caused by the existence of holes in the ozone layer. They point out that a major portion of Australian skin cancers occurs because the country lies near the equator, where sunlight and ultraviolet radiation are more intense.

Scientists know that repeated exposure to ultraviolet radiation causes skin cancer. Skin cancer rates are highest for people who frequently work or play outside.

Australia's location near the Antarctic hole,

however, is a very real problem. On several occasions between 1988 and 1990, Australian scientists detected large patches of ozone-depleted air that had apparently traveled from the South Pole. When the patches were over Australia, ultraviolet radiation soared to levels more than twenty times higher than normal. Australian television stations now broadcast daily ultraviolet radiation warnings. When the rays reach dangerous levels, authorities advise people to stay indoors as much as possible.

In addition, studies show that skin cancer is increasing annually in Australia. Several researchers have studied this problem, including University of Miami chemistry professor David E. Fisher. "In Tasmania," Fisher says, "the Australian state closest to the Antarctic ozone hole, the incidence of [malignant] melanomas has doubled in the past ten years."

Other ultraviolet damage

Ultraviolet rays can cause other health problems besides skin cancer. High doses can impair the immune systems of mammals, including human beings. The immune system is the body's main defense against disease. If a person's immune system is not working properly, he or she is more likely to contract cold sores, skin infections, and skin diseases.

Ultraviolet light is also a major cause of eye cataracts. These are patches of opaque, or light-blocking, tissue that form in the eyes and can lead to partial or complete blindness. Laboratory experiments have shown that ultraviolet rays are 250 times more likely than regular light to cause cataracts. According to the EPA, if the present rate of ozone depletion continues for the next forty years, the increased ultraviolet radiation could cause as many as ten million extra cases of cataracts in the United States alone.

Animals and people are not the only creatures harmed by ultraviolet rays. Many microorganisms that produce nutrients in the soil die from overexposure to these rays. The microorganisms are extremely sensitive to changes in the intensity of the rays. So continued upper-level ozone depletion could potentially decrease soil fertility, especially in tropical regions where ultraviolet radiation is naturally more intense.

The threat to plant life

Plants, too, can suffer from increased doses of ultraviolet rays. According to researcher Robert C. Worrest of the EPA, scientists have tested about two hundred species of land plants—mostly food crops—for sensitivity to ultraviolet light. Worrest says:

> The most sensitive plant groups include crops related to peas and beans, melons, mustard and cabbage. . . . In general, ultraviolet radiation causes reduced leaf and stem growth [and] lower total dry weight. . . . In addition to other factors, increased levels of [ultraviolet] radiation may reduce the quality of crop yield. . . . Existing data also suggest that increased radiation will modify the distribution and abundance of plants.

Worrest and his colleagues stress that scientific knowledge in this area is still limited and that thousands of other plant species have not yet been tested. They call for more worldwide studies. More information, they say, will help prevent potentially serious crop losses that might occur if ozone depletion continues.

Ocean plants can also be damaged by too much exposure to ultraviolet rays. Some experiments suggest that tiny plants like algae and plankton, which move by the trillions through the seas, are extremely sensitive to ultraviolet light. These species are important for two reasons. First,

through the process of photosynthesis, they produce a significant amount of the oxygen in the earth's atmosphere. Second, they occupy the base of nature's food chain. Microscopic animals eat these tiny plants. Larger animals then eat these microscopic animals, and the process continues upward through the animal kingdom. Human beings are at the top of the food chain. Many scientists believe that increases in ultraviolet radiation caused by upper-level ozone depletion could significantly reduce algae and plankton found in the oceans.

The same researchers admit that they still do not have enough information to predict how many of these plants and animals might be lost. Some scientists think that there might not be any loss at all. They point out that ozone depletion is happening gradually, over the course of years, while the life cycle of plankton spans only a few days. So the tiny plants might slowly be able to adapt themselves to higher ultraviolet levels. In that case, changes in the workings of the food chain might be minimal. All the researchers agree that long-term studies are needed before anyone can say with absolute certainty what ozone depletion will do to ocean plants and animals.

Searching for the culprit

All these potentially harmful effects of ultraviolet radiation are a cause for concern to scientists and government leaders around the world. Researchers work to discover the reasons why this ozone depletion occurs. They also search for ways to slow and eventually halt the destruction of upper-level ozone. This research is relatively new. Until a few years ago, no one knew about the Antarctic ozone hole or even that a decrease of upper-level ozone was possible.

The discovery of ozone depletion came in the early 1970s. Before that, scientists paid little attention to the ozone layer. This was mainly because they did not yet thoroughly understand the chemistry of the upper atmosphere. In 1970, two atmospheric scientists independently made the same suggestion. Paul Crutzen, then a researcher at the University of Stockholm in Sweden, and Harold Johnson, a researcher at the University of

For a while, scientists feared supersonic aircraft like this one would damage upper-level ozone. Now, though, only a few remain in use, and they have not noticeably affected the ozone layer.

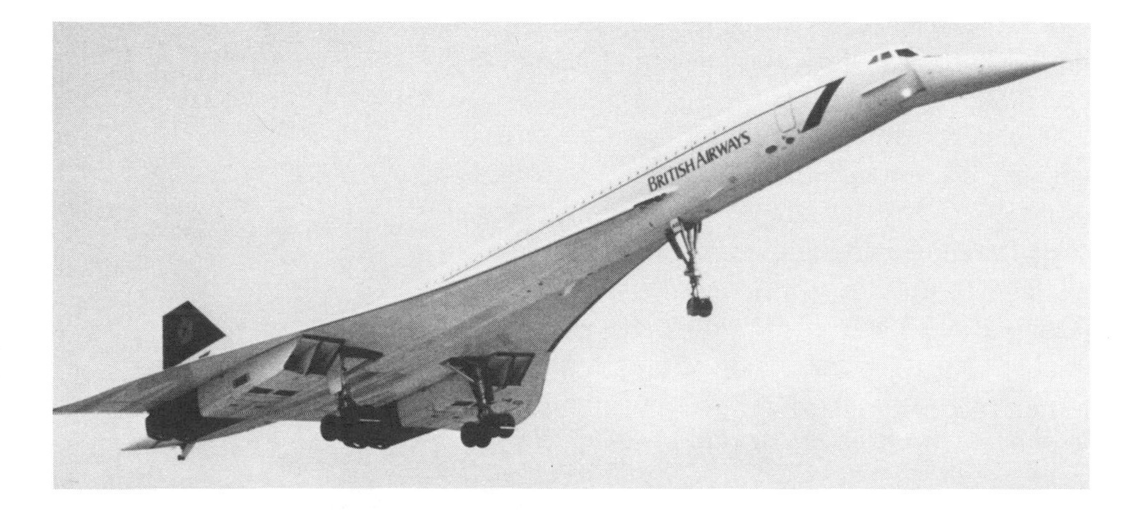

California at Berkeley, each suggested that exhausts from high-flying airplanes would introduce large amounts of pollutants into the atmosphere. These pollutants, the scientists said, might seriously damage the ozone layer.

Crutzen and Johnson's idea generated a great deal of publicity because, at the time, several countries were considering building fleets of supersonic aircraft that would fly high in the atmosphere. The United States, France, and Great Britain all proposed launching these passenger-carrying craft. The United States eventually canceled its plans. This was partly due to fear of damaging the upper ozone layer. Meanwhile, both France and Great Britain built their supersonic fleets. Now, only a few of the aircraft are in use, and they have not noticeably affected upper-level ozone.

Some scientists believe that exhausts from the space shuttle affect the ozone layer.

The next step in the search for the ozone-destroying culprit came in 1974. Richard Stolarski and Ralph Cicerone, both of the University of Michigan, did a study for the National Aeronautics and Space Administration (NASA). They wanted to see if exhaust from the space shuttle would affect the ozone layer. They found that one element in the exhaust—chlorine—did affect the ozone layer. Further studies showed that a single chlorine atom will likely destroy between 10,000 and 100,000 ozone molecules. Chlorine's potential for eliminating ozone is enormous.

Destruction in a spray can

The new knowledge of chlorine-ozone reactions disturbed many scientists. They pointed out that people had unknowingly been throwing chlorine into the atmosphere for years. Two of these concerned researchers were Mario Molina and F. Sherwood Rowland of the University of California at Irvine. They argued in a 1974 paper that manu-

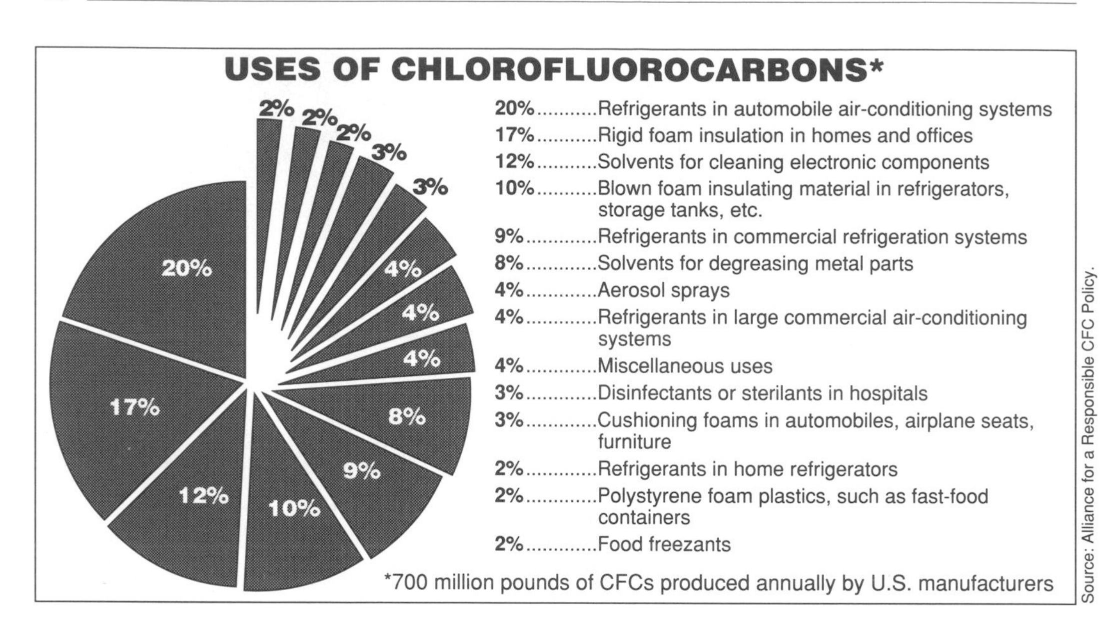

factured substances called chlorofluorocarbons (CFCs) could pose a threat to the ozone layer.

As the name suggests, CFCs are composed of atoms of chlorine, fluorine, and carbon. Science writer Michael D. Lemonick sums up the characteristics of CFCs:

> When they were first synthesized [created in the lab] in the late 1920s, [CFCs] seemed too good to be true. These remarkable chemicals . . . are nontoxic and inert, meaning that they do not combine easily with other substances. . . . CFCs are perfect as coolants in refrigerators and propellant gases for spray cans. Since CFCs are good insulators, they are standard ingredients in plastic-foam materials like Styrofoam. Best of all, the most commonly used CFCs are simple, and therefore cheap, to manufacture.

But Molina and Rowland pointed out that CFCs have a major drawback, one that no one had ever before considered. Although their molecules are sturdy under normal conditions, they break apart easily in the presence of intense ultraviolet radiation. Molina and Rowland suggested that many of the CFCs released through human activity slowly drift upward into the stratosphere. There, near the top of the ozone layer, they encounter concentrated doses of ultraviolet light. The CFCs then break apart, and the freed chlorine begins destroying ozone.

Rowland pointed out that human production of CFCs in the mid-1970s already totaled more than one million tons per year. And that figure was doubling every five to seven years. Claiming that CFCs could eventually cause serious damage to the ozone layer, he called for a ban on CFC propellants used in spray cans.

Ozone depletion on the back burner

Other scientists in the United States and around the world quickly confirmed Molina and Rowland's calculations. The public became concerned about CFC use, and some U.S. legislators even considered banning the chemicals. However, CFC manufacturers, including DuPont—the biggest CFC producer in the country—asked for more time to study the problem. Industry spokespeople agreed that chlorine could destroy ozone. But, they insisted, there was no way to tell just how concentrated the CFCs in the ozone layer had become. Perhaps very few CFCs made it to that altitude, they suggested. And besides, people had used CFCs for decades, and there was no evidence that the ozone layer had been impaired at all, they said.

Despite the lack of solid evidence, the EPA decided to play it safe in 1978. The agency banned the use of CFCs as propellants in spray cans. Companies found other, safer gases for propellants or switched to pump sprays. Many people breathed a sigh of relief, assuming that the CFC problem had been eliminated.

But it was not. The CFCs used in sprays were only partly responsible for the ozone depletion.

Though the EPA banned CFC propellants in spray cans, the CFC problem has not been eliminated. CFCs are still used for such things as coolants for refrigerators and air conditioners, Styrofoam, and industrial cleaners.

People still used CFCs in many other ways: coolants for refrigerators and air conditioners, foam for insulation and packaging, industrial cleaners, and substances for sterilizing medical instruments. Also, people in other countries still used CFC sprays. The U.S. ban on CFCs in sprays did little to keep the chemicals from building up in the atmosphere.

In spite of the buildup of these chemicals, there was still no clear-cut evidence that they actually posed a serious threat to the ozone layer. Some scientists said their colleagues had exaggerated the harmful potential of CFCs. Researchers at the National Academy of Sciences in Washington, D.C., admitted that they were unsure about the

amount of ozone that might be depleted. They issued widely varying estimates. Government officials also began to take the CFC threat more lightly. In 1981, EPA administrator Anne Gorsuch stated that ozone depletion was not a serious problem.

A hole in the sky

In 1984, a single event brought the subject of ozone depletion back to the world's attention. Using satellites, balloons, and ground-based instruments, British scientists discovered a hole in the ozone layer. Centered over Antarctica, an area of the stratosphere hundreds of miles wide was depleted of at least two-thirds of its ozone. Soon, several nations sent specially equipped airplanes through the Antarctic stratosphere in an attempt to find reasons for the disturbing ozone losses.

Within months, the researchers discovered that a previously unknown type of chemistry was at work over the South Pole. In the frigid air of the polar stratosphere—where temperatures often reach -110° F—a huge vortex, or circular whirlwind, spins around and around. Within the vortex are strange clouds, which scientists call polar stratospheric clouds. These cold clouds have unusual physical properties that speed the separation of chlorine from various chlorine compounds. Sure enough, when the scientists examined the polar clouds, they found the chlorine concentration to be hundreds of times greater than normal levels. There was no longer any doubt. The threat of ozone depletion was real.

After the discovery of the Antarctic ozone hole, ozone-related research expanded rapidly. In 1986, 130 scientists from around the world organized a planetwide study of the ozone-depletion phenomenon. They set out to verify the details of the cold-air chemistry over Antarctica and also to determine if factors other than CFCs were contributing to the problem. For instance, some thought that the satellites and other measuring instruments might be malfunctioning and reporting misleading data. Others theorized that sunspots, dark patches that appeared on the sun, might be emitting excess radiation that affected the polar air.

Volcanic eruptions

Still another possibility was that volcanic eruptions had something to do with the diminished ozone levels in Antarctica. Some researchers pointed out that the Antarctic volcano Mount Erebus has been erupting almost continuously for more than one hundred years. Science writer Rogelio Maduro explains:

> In 1983, samples were taken of the gases being blown into the atmosphere by Mt. Erebus, indicating that more than 1,000 tons of chlorine were being outgassed [given off] daily. Given the high altitude of the volcano and the extremely dry conditions in Antarctica, which prevent the chlorine gases from being washed to the ground by rain, volcanologists estimate that a very large percentage of these chlorine gases must reach the stratosphere.

In 1988, the scientists announced the results of the planetwide study. They found that the ozone layer was definitely eroding, mainly because of large amounts of chlorine in the stratosphere. Exactly where this chlorine was coming from was still a matter of debate. A few researchers still felt that Mount Erebus and also a Mexican volcano that had erupted in 1982 were the major sources of the chlorine. Others pointed out that the natural evaporation of sea water carries millions of tons of salt per year into the atmosphere, and chlorine is an ingredient of salt. As Rogelio Maduro put it, these and other natural sources "pump hundreds of millions of tons more chlorine into the atmosphere every year than is put there by all man-made CFCs."

But many of the scientists did not believe that the bulk of the chlorine in the ozone layer came from volcanoes and evaporated salts. They agreed that these sources injected a great deal of chlorine into the atmosphere, much more than CFCs did. But they were not convinced that enough of this chlorine traveled high into the stratosphere. The view of a majority of the researchers was that CFCs *did* travel that high and accounted for more of the excess chlorine at greater altitudes.

The researchers further concluded that there had been an overall 2 percent drop in upper-level ozone in the northern hemisphere (above the equator) between 1978 and 1987. The ozone de-

Volcanoes pump millions of tons of chlorine into the atmosphere. Scientists disagree, however, about whether the chlorine travels high enough into the stratosphere to damage the ozone layer.

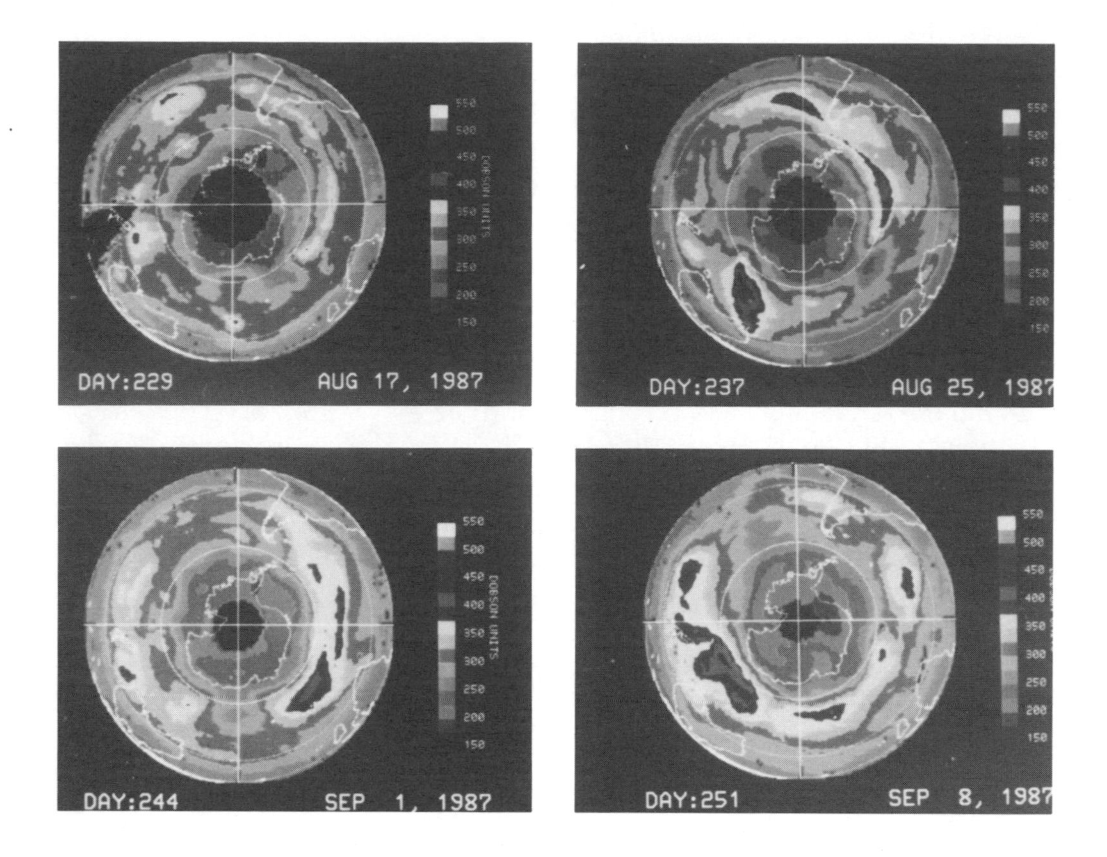

Scientists study computer-generated models of the Antarctic polar region to monitor changing ozone levels.

pletion in the southern hemisphere during that same period was 3 percent. Depletion within the Antarctic hole itself varied from about 60 percent to 70 percent. And the rate of depletion appeared to be increasing. During a six-week period in 1987, 90 percent of the ozone in the center of the hole disappeared. The scientists also used computers to build "models" of ozone depletion to help predict future losses of the gas. The models suggested that the losses would continue.

To make matters worse, in 1989, NASA scientists discovered a similar, though much smaller, area of ozone depletion over the Arctic, the north polar region. And researchers began to detect patches of ozone-depleted air floating over Australia and other areas of the southern hemisphere.

Some researchers remain unconvinced about the seriousness of upper-level ozone depletion. They acknowledge that the Antarctic ozone hole exists, but they believe it to be a natural phenomenon. For instance, atmospheric scientist Gordon Dobson and NASA scientist Mark R. Schoeberl think the hole is related to changes in weather. They contend that the hole is peculiar to Antarctica and not part of a global trend toward ozone depletion.

But these views represent a minority within the scientific community. Most scientists now believe that evidence for worldwide stratospheric ozone depletion is overwhelming. Even DuPont, the chief producer of CFCs, has accepted this data. The company originally argued that "scientific evidence does not point to the need for dramatic CFC . . . reductions." But in March 1988, DuPont announced that it would phase out CFCs by the end of the century and called for a complete ban on the chemicals. Commenting in 1989 on DuPont's turnaround, physicist David E. Fisher observed, "Their decision . . . [came] none too soon. The latest data showed that global ozone concentrations had dropped several percent over the last decade, which was even faster than the computer models had predicted."

Scientists had found that, apparently, there was too little ozone where it was needed most. And, according to some researchers, there is even more to worry about. Ozone might be involved in a trend that may be causing global temperatures to increase. The effects of these higher temperatures could be disastrous.

4

Ozone and Global Warming

MANY SCIENTISTS believe that low-level ozone is presently contributing to a slow but steady increase in the temperature of the earth's atmosphere. These scientists say that this global warming could eventually lead to significant and harmful changes in the planet's climate.

A global greenhouse?

Ozone may contribute to global warming by trapping heat in the lower atmosphere. It does this in the same way that a glass greenhouse captures and holds heat. When sunlight shines on a greenhouse, the rays pass straight through the transparent glass. The air molecules inside the closed structure absorb the heat from the solar rays. The warmed air, unable to pass back through the glass, remains trapped inside. So as long as the sun is shining, the air inside the greenhouse gets warmer and warmer.

Low-level ozone in the atmosphere works similarly. The ozone molecules absorb heat from incoming sunlight. They then push the heat back toward the earth's surface, causing heat to build up in the lower atmosphere. This process is known as the greenhouse effect.

(opposite page) Some scientists believe that too much ozone will affect the atmosphere surrounding the earth, pictured here in a photo from the Apollo 8 spacecraft.

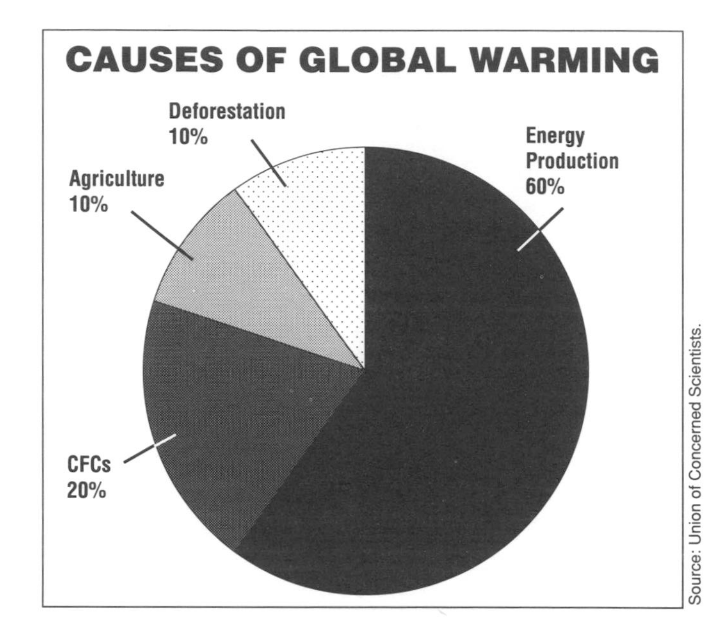

Because ozone molecules absorb heat, helping to create this effect, scientists refer to ozone as a "greenhouse gas." It is not the only greenhouse gas. There are several others, including water vapor, carbon dioxide, methane, nitrogen oxides, and CFCs. When all of these gases are present in the atmosphere, the warming trend increases, according to some scientists. This is because ozone and these other gases trap heat very efficiently.

The amount of low-level ozone in the atmosphere has been increasing since the 1950s. Scientists have shown that there is presently 100 to 200 percent more tropospheric ozone in the atmosphere than there was a century ago. Considering ozone's strong heat-trapping powers and its increased concentration in the troposphere, many researchers believe that ozone may play an even larger role in the warming of the earth's atmosphere in the next century.

Some scientists believe this warming has already begun and that it is likely to get worse. One such scientist is James Hansen of NASA's Goddard Institute for Space Studies in New York. In 1988, Hansen told a group of U.S. senators, "With 99 percent confidence . . . the greenhouse effect has been detected [in the atmosphere] and is changing our climate now." Other scientists say evidence collected so far does not support the idea that global warming is a trend that people have to worry about. They say the atmosphere's recent warming is the result of natural climatic changes. One such scientist is Reid Bryson, a founder of the Institute for Environmental Studies at the University of Wisconsin. In 1989, Bryson said, "The very clear statements that have been made [by Hansen] that the greenhouse warming is here already and that the globe will be . . . warmer in fifty years cannot be accepted."

The possible consequences

At present, no one is absolutely certain how dangerous global warming might be. Nevertheless, some scientists have tried to predict what global warming could mean to the earth and its inhabitants. These scientists say the climate changes that would accompany global warming would

transform life as we know it today.

The EPA conducted a series of studies in the late 1980s on the subject of global warming. The aim of these studies was to help both the government and the public understand the environmental changes that could result if global warming occurred.

One EPA study examined what would happen if average global temperatures rose five degrees. The conclusion was that less rain and snow would fall in many parts of the United States. This, in turn, would mean less water for growing crops. This would be especially hard on certain crops. Corn, for example, needs a minimum amount of rain at a certain time in its growing season, and wheat requires a lot of groundwater from melting snow. In addition, the EPA report stated that a

Wheat crops in the United States could be damaged by global warming as the amount of precipitation decreases.

five-degree temperature increase would lead to greater evaporation of lake and river water. Colorado River runoff, for example, would fall by at least 10 percent. This would, in turn, affect food production in large sections of the Southwest that rely on the river's runoff for irrigation.

Another EPA study looked at possible changes in sea level. If global temperatures rise, this study said, at least part of the polar ice caps would melt. This would cause sea levels around the world to rise. According to the EPA, a rise of just three feet would flood most U.S. coastlines. A five-foot rise in sea levels would mean severe flooding for coastal cities like New York and Los Angeles, the EPA predicted. Some parts of the world would fare even worse. Low-lying countries like the Netherlands and Bangladesh would lose large chunks of their land to flooding.

Although global warming is still being debated, many experts and organizations, including the EPA, have called for action now. They propose steps toward reducing human production of greenhouse gases, especially ozone, carbon dioxide, and CFCs. They urge that these steps be part of an overall environmental strategy. Those concerned with the issue say, in short, that people need to do something about ozone.

BOYC
EXXON
CONSE
N
EXPL
LOVE
FLOW
CEO
UTER
STOP
OZ ONE
NO-
ZONE

5

What People Are Doing About Ozone

(opposite page) Environmentalists gather on Earth Day 1989 to protest practices that they believe threaten the earth.

Efforts by various governments and organizations to do something about ozone-related problems are already underway. These efforts fall into two general categories: 1) attempts to eliminate low-level ozone pollution and 2) attempts to slow the depletion of the upper ozone layer.

Fighting ozone pollution

The battle against low-level ozone pollution in the United States began as a battle against air pollution in general. When Congress established the EPA in 1970, the new organization immediately tackled the problem of air pollution. EPA officials set standards for various pollutants, including ozone. The goal was to have all parts of the country meet these standards by 1976.

Many areas of the United States did not comply with the EPA's standards by 1976. So Congress extended the deadline to 1982 for most areas and to 1987 for areas with the worst levels of ozone and other pollutants.

These efforts to control air pollution centered on

Although air pollution controls in the 1970s got rid of many of the pollutants released by cars and trucks, ozone levels remain high in many U.S. cities. In 1988, ninety-six cities and counties failed to meet EPA standards.

reducing ozone-producing waste materials emitted into the air by factories and motor vehicles. The idea was to "scrub" the pollutants from the exhausts by using filtering devices. The most familiar example is the catalytic converter, used to convert automobile exhaust into mostly harmless substances. As exhausts pass through the converter, chemical reactions remove many of the pollutants. Larger, more complex versions of these converters are found in the smokestacks of factories and oil refineries. Each year, the devices absorb millions of tons of pollutants that would otherwise flow into the atmosphere and produce ozone.

The use of catalytic converters and other filtering devices seemed to help for a while. Levels of some pollutants fell in the late 1970s and early 1980s. At the same time, tropospheric ozone levels decreased slightly in urban and industrial areas like Los Angeles and Detroit.

Ozone refuses to go away

While low-level ozone in the cities decreased, ozone levels increased in many rural areas. This situation at first surprised and confused scientists. But they soon found that nitrogen oxides were causing the problem. The catalytic converters in use did not remove nitrogen oxides from the exhausts, and winds carried these chemicals far out into the countryside. There, the gases mixed with hydrocarbons given off by trees. The addition of sunlight created plenty of ozone.

What people needed were catalytic converters that also removed nitrogen oxides from exhausts. Actually, such converters have existed for many years. But they are much more expensive than standard converters. The auto industry was reluctant to install the more costly devices. Industry spokespeople argued that regular converters had already substantially reduced air pollution, making the nitrogen oxide converters unnecessary.

In the 1980s, despite the widespread use of catalytic converters and other antipollution measures, air pollution increased again. And concentrations of low-level ozone grew faster than those of other pollutants. This was mainly because of a huge increase in the number of cars and trucks in the United States. The number of these vehicles increased by more than twenty-five million between 1980 and 1987. More vehicles burn more fuel. U.S. cars and trucks burned about three billion more gallons of gasoline per year in the mid-1980s than in the mid-1970s. All this extra fuel burning caused a rise in air contamination, in-

Smog hovers over the World Trade Center in New York City. New York City has until the year 2010 to comply with standards set out in the Clean Air Act of 1990.

cluding ozone pollution.

On June 12, 1989, President George Bush spoke to the nation's governors about a new attack on air pollution. "Too many Americans breathe dirty air," he stated, urging every citizen to join the fight against air pollution. The president's plan, a version of a bill debated in Congress for several years, called for stricter controls on exhausts from factory smokestacks and car tailpipes. He also proposed that more gas stations use pumps that prevent gasoline fumes from floating into the atmosphere. The plan called on gasoline companies to make a product that would be less likely to evaporate into the air. In addition, the president talked about the possibilities of alcohol fuels for cars and trucks.

Meeting ozone standards

This new plan also set goals for meeting ozone standards. All cities would have to meet the existing EPA low-level ozone standard by the year 2000. The exceptions would be the three cities with the worst smog problems—Los Angeles, New York, and Houston. These cities would have until the year 2010 to comply with the ozone standards.

President Bush and Congress finally agreed on the plan in 1990. On November 15, 1990, Bush signed into law what has been hailed as the most comprehensive air pollution measure introduced so far in the United States. The new Clean Air Act gives industry until the turn of the century to cut emissions of pollutants that cause smog and deplete the protective ozone layer. This includes a complete phaseout of CFCs. The bill also requires manufacturers to produce cars and fuels that pollute less or not at all.

Some officials feel the bill is not tough enough on ozone. California officials, for example, said they had hoped for even stricter standards on pollution created by cars, small construction equipment, and farm equipment. But other people praised the bill, saying it was a step in the right direction. Those concerned about low-level ozone were especially pleased that the bill required car manufacturers to design vehicles that burn fuels other than gasoline. One of the most frequently mentioned of these alternative fuels is alcohol.

The switch to vehicles that burn alcohol instead of gasoline would take place mainly in ar-

In 1979, a gas station attendant pumped gasohol for the first time in the state of New York. Gasohol burns cleaner than gasoline, and therefore causes less ozone pollution. However, alcohol fuels release formaldehyde, a known carcinogen, into the air.

eas like Los Angeles, where ozone smog is severe. Alcohol burns much cleaner than gasoline and releases fewer pollutants into the air. So cars that burn alcohol produce fewer chemicals that produce ozone.

The idea of using alcohol as a fuel is not new. Alcohol-burning cars were designed by several companies in the 1960s. And Brazil began large-scale use of such cars more than ten years ago. In 1990, Brazil had more than 2.5 million alcohol-powered vehicles on the road. The country also uses alcohol to power several industrial plants. As a result of these practices, low-level ozone and smog caused by cars in many Brazilian cities significantly decreased in the 1980s. Brazil's new fuel consists of two types of alcohol—methanol and ethanol. Both are made from wood and vari-

© Christian/Rothco. Reprinted with permission.

ous crops and plants, especially sugarcane, which is commonly grown in Brazil.

In the United States, some people have been using alcohol fuels since the 1970s. They buy a product called gasohol, a mixture of ethanol and gasoline. Gasohol can be used in most car and truck engines, but it is presently available at only a limited number of service stations. American vehicles that run strictly on alcohol are even harder to find. This situation may soon change. Both the Ford and Volkswagen companies now manufacture vehicles that run on ethanol.

Truckloads of sugarcane arrive at this plant in Brazil. The sugarcane will be used to make alcohol and will then be mixed with gasoline to form gasohol. Using gasohol is one means of extending limited fuel reserves.

Using alcohol as a fuel does have its drawbacks. Although alcohol fuels like ethanol burn cleaner than gasoline and therefore cause less ozone pollution, they release formaldehyde into the air. Formaldehyde is a chemical known to cause cancer. Methanol and ethanol also cause more wear and tear on car engines and make cold engines harder to start. Car designers are trying to eliminate these drawbacks of alcohol fuels.

Reversing the warming trend

The switch to alcohol and other cleaner-burning fuels is also welcomed by environmentalists as a way of slowing global warming. Such fuels produce fewer greenhouse gases. Therefore, they could help reverse the warming trend that some scientists believe is steadily worsening.

The use of alternative fuels is one of the steps proposed by the EPA to slow the greenhouse effect. The agency also proposed increasing the price of oil and other fossil fuels. The agency hopes this will encourage people to use less of these polluting substances and change to cleaner alternatives. The EPA has also suggested expanding the use of energy alternatives like solar and nuclear power. Neither produces low-level ozone or any other greenhouse gases.

Reducing levels of CFCs is also part of the worldwide effort to deal with the other major ozone-related problem—depletion of the upper ozone layer. After the discovery of the Antarctic ozone hole, people began to seriously consider limiting CFC production. In 1985, the United Nations brought together representatives from twenty-one countries to talk about CFC restrictions.

The Soviet Union, Japan, and twelve European nations wanted to freeze CFC production at existing levels. They argued that getting rid of the chemicals altogether would be a mistake for economic reasons. CFCs are simple and cheap to produce. Substitutes, they said, would be costly to develop and manufacture, and that would cause prices of refrigerators, insulation, and other products to climb.

No specific agreement

The United States, Canada, and the Scandinavian countries called for a complete phaseout of CFCs by the year 2000. They argued that the health risks, including increased skin cancers, were more pressing than economic concerns. Also, they pointed out that CFCs often take many years to reach the stratosphere. That means that CFCs released in 1985 still could be doing damage to upper-level ozone in 2025. Therefore, the sooner these materials were eliminated, the better. The meeting ended without any specific agreement.

Congressman Perry Bullard's 1975 campaign to outlaw aerosol propellants helped pave the way for their eventual ban.

In the meantime, Secretary of the Interior Donald Hodel warned skeptical U.S. government officials of the alternatives to dealing with ozone depletion. Everyone, he said, could remain indoors all the time or always wear hats, sunglasses, and tanning lotion to guard against increased levels of ultraviolet radiation.

The countries met again in 1987 in Montreal, Canada. This time, they reached an agreement on

CFC reduction. On September 16, 1987, twenty-four countries signed the Montreal Protocol. Those signing included all the world's major producers of CFCs. By 1989, another fifteen countries had signed the agreement. The nations agreed to reduce production and use of CFCs by 20 percent by 1993 and by another 30 percent by 1998.

The Montreal Protocol was an important step in helping end the ozone-depletion threat. But further research in 1988 and 1989 showed that the depletion problem was even worse than previously thought. Scientists and government officials realized that the Montreal agreement had not gone far enough. In March 1989, the twelve European nations that had opposed the U.S. phaseout proposal in 1985 changed their position. They voted to halt all production of CFCs by the year 2000. Just two months later, in May 1989, eighty-one nations met in Helsinki, Finland. They adopted a declaration for a complete ban on CFCs by the end of the century.

Representatives of the international business community have accepted the coming ban as a necessary action to save the environment and eliminate human health risks. DuPont's voluntary cooperation in the phaseout is a prominent example. But many businesses are bracing themselves for the changeover. Douglas G. Cogan, of the Investor Responsibility Research Center in Washington, D.C., explains that people have come to take the benefits of inexpensive CFCs for granted. He predicts, "Virtually every American household, most of the nation's transportation fleet, and 375,000 business locations will experience withdrawal symptoms as the nation weans itself of its daily dependence on these chemicals."

The CFC dilemma

Cogan points out that politicians could choose to eliminate CFCs from the marketplace much sooner than the year 2000. But that would create short-term economic problems. Because of these considerations, the CFC issue has put government officials around the world in a dilemma. As Cogan explains:

> A consensus is emerging . . . that we must prepare for a world without CFCs. The key question for policymakers is whether we should rush to eliminate these chemicals from production—imposing greater hardships on ourselves and on our economy or wait for more alternatives to emerge, thereby imposing more serious environmental and health consequences on future generations.

As politicians wrestle with the issue of regulating CFCs, industry has tackled the challenge of inventing safer alternatives to these chemicals. Chemical companies are already testing several promising CFC substitutes. For example, the Dow Chemical Company has stopped using CFC-11 in the production of foam packaging. The company now uses another chemical that is twenty times less harmful to the ozone layer than CFC-11. And DuPont plans to use a substance called HFC-134A as a coolant in car air conditioners and

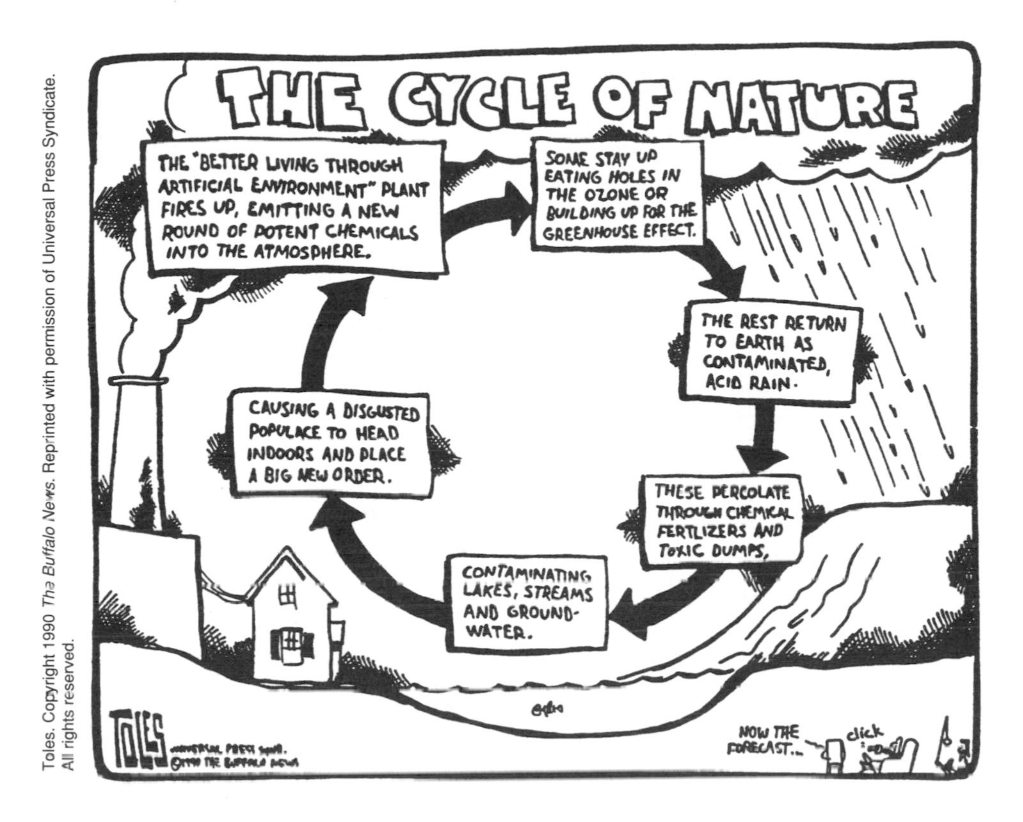

other products. HFC-134A contains no chlorine. So even if it reaches the stratosphere and breaks down under ultraviolet rays, it will not harm the ozone layer. Industry spokespeople say that HFC-134A should reach the commercial market by 1993.

Other companies are working on CFC substitutes. AT&T, along with Petroferm, a Florida chemical producer, came up with a substance made from orange peels to replace CFC-13. And Union Carbide discovered a new method for producing foam for bedding and upholstered furniture. The procedure requires no CFCs.

Efforts to curb low-level ozone pollution and halt upper-level ozone depletion have only just begun. These ozone-related problems became readily apparent only in the second half of the twentieth century. Considering how little time sci-

A flag of the earth waves over this crowd gathered at the nation's capital. Over 100,000 attended a rally in 1990 to celebrate the twentieth anniversary of the first Earth Day.

entists have had to deal with these problems, concerned governments and individuals have responded relatively quickly. This reflects not only the impressive advances of modern science but also the dedication of thousands of people around the world to ensuring the well-being of the planet.

No one can say how long it will take to bring ozone back into the balance it once maintained in nature. It will be a huge job. No single scientific breakthrough or legislative action will be enough. Large industries, small businesses, and individual citizens must all share in the task. This will sometimes involve throwing out old products and learning to use new, safer ones. Many people will have to alter their life-styles by conserving fuel and switching to cleaner and more efficient technologies.

Most important, ordinary people everywhere

will need to become more conscious of the environment, more aware of how it can be damaged, and more educated about what is required to repair the damage already done. As Jack Fishman and Robert Kalish, authors of *Global Alert: The Ozone Pollution Crisis*, put it, "Despite national borders and differences among peoples, we are all residents of the same smoke-filled room—the planet earth . . . [and] its continued existence is everyone's responsibility."

Glossary

algae: Tiny ocean plants.

atoms: The microscopic building blocks that make up all matter.

catalytic converter: A device used in motor vehicles to remove pollutants from the exhaust.

cataracts: Small, light-blocking patches that form in the eyes of animals and people.

chlorofluorocarbons (CFCs): Substances composed of atoms of chlorine, fluorine, and carbon; used as coolants and in insulation and other products.

early autumn syndrome: A condition in which ozone and other pollutants cause the leaves of trees to die in summer instead of autumn.

emphysema: Lung disease in which the victim cannot get enough oxygen from normal breathing.

ethanol: A form of alcohol sometimes used as a fuel.

formaldehyde: A cancer-causing chemical released by the burning of alcohol.

gas chromatography: A method of measuring ozone concentrations. A long tube holds several filters that remove unwanted gases from an air sample; the remaining gases are burned to determine their proportions in the sample.

global warming: The increase in atmospheric temperature due to the buildup of carbon dioxide and other substances in the air.

greenhouse effect: The process by which air is warmed as heat is trapped by glass, gases, or other substances.

greenhouse gases: Gases that easily absorb heat and create the greenhouse effect; these gases include ozone, carbon dioxide, and CFCs.

hydrocarbons: Substances composed of atoms of hydrogen and carbon, commonly emitted as waste pollutants when fuel is burned and produced by plants during photosynthesis.

iron oxide: Rust formed when iron reacts with oxygen or ozone.

malignant melanoma: The most dangerous type of skin cancer.

melanin: A pigment in the skin that blocks ultraviolet rays; dark-skinned people have more melanin than fair-skinned people.

methane: A gas composed of atoms of carbon and hydrogen, given off during the decay of dead plant and animal tissue.

methanol: A form of alcohol sometimes used as a fuel.

microorganisms: Microscopic plants and animals.

molecule: A combination of two or more atoms.

nitrogen oxides: Chemical compounds containing atoms of nitrogen and oxygen; an example is nitric oxide, a pollutant emitted in car exhausts.

oxidation: The process by which oxygen and ozone chemically combine with other materials, as in the formation of rust.

oxidizer: A substance, such as ozone, that easily combines with other materials.

ozone: A gas made up of oxygen atoms; each ozone molecule contains three atoms of oxygen.

ozone depletion: The destruction of the stratospheric ozone layer by chlorine and other substances.

ozone layer: The band of ozone circling the earth in the stratosphere; the layer absorbs much of the dangerous ultraviolet radiation that would otherwise reach the earth's surface.

photosynthesis: The process by which green plants combine carbon dioxide from the air, nutrients from the soil, and sunlight to create energy; oxygen is released as a by-product.

plankton: Tiny ocean plants.

polar stratospheric clouds: Very cold clouds formed in the stratosphere above the South Pole; they speed up the depletion of ozone by chlorine.

propellant: A chemical used to force liquids out of a spray can.

smog: Air pollution composed of primary pollutants, such as sulfur dioxide and soot; the word is a combination of the words *smoke* and *fog*.

stratosphere: The layer of the atmosphere located between twelve and twenty-four miles above the earth's surface.

sulfur dioxide: A primary pollutant made up of molecules containing atoms of sulfur and oxygen.

troposphere: The layer of the earth's atmosphere located between the surface and an altitude of about six to nine miles.

ultraviolet light: Rays from the sun that can be dangerous to organisms.

Suggestions for Further Reading

Alice Bredin, "On Ill Health and Air Pollution," *The New York Times Good Health Magazine*, October 7, 1990.

C. Claiborne, "Can We Fix the Ozone Hole?" *National Wildlife*, February/March 1990.

Discover, "Leaves of Gas," February 1989.

David E. Fisher, *Fire and Ice: The Greenhouse Effect, Ozone Depletion and Nuclear Winter*. New York: Harper & Row, 1990.

S. Flamsteed, "Ozone Ups and Downs," *Discover*, January 1990.

Michael D. Lemonick, "Deadly Danger in a Spray Can," *Time*, January 2, 1989.

Newsweek, "And Now the Return of the Killer Trees?" June 5, 1989.

Works Consulted

J.E. Basu, "Why No One's Safe," *American Health*, September 1989.

Consumer Reports, "Can We Repair the Sky?" May 1989.

Jack Fishman and Robert Kalish, *Global Alert: The Ozone Pollution Crisis*. New York: Plenum Press, 1990.

The Futurist, "Economics of Preserving the Ozone Layer," January/February 1989.

T.A. Heppenheimer, "Keep Your Cool," *Reason*, January 1990.

B. Hogan, "1989 Antarctic Ozone Loss Severe," *The Conservationist*, May/June 1990.

Robin Russell Jones and Tom Wigley, eds., *Ozone Depletion: Health and Environmental Consequences*. New York: John Wiley & Sons, 1989.

R.A. Kerr, "Ozone Destruction Closer to Home," *Science*, March 16, 1990.

Rogelio Maduro, "The Myth Behind the Ozone Hole Scare," *21st Century Science & Technology*, July/August 1989.

R. Monastersky, "Antarctic Ozone Hole Unexpectedly Severe," *Science News*, February 25, 1989.

R. Monastersky, "CFC Replacements: Better but Not Ideal," *Science News*, April 7, 1990.

R. Monastersky, "The Two Faces of Ozone," *Science News*, September 2, 1989.

D. Moreau, "Sherry Rowland Linked a Nontoxic Chemical

to an Environmental Catastrophe," *Changing Times*, January 1990.

J. Raloff, "Ozone: Indoors May Offer Little Protection," *Science News*, September 23, 1989.

Science News, "Europe to Ban CFCs by 2000," March 11, 1989.

Science News, "Ozone Hole Threatens Polar Plankton," October 28, 1989.

Science News, "Potential Replacement for Ozone Killer, " February 11, 1989.

S.F. Singer, "My Adventures in the Ozone Layer," *National Review*, June 30, 1989.

Time, "A Baffling Ozone Policy," May 21, 1990.

Melinda Warren and Kenneth Chilton, "Cleaning the Air of Ozone," *Society*, vol. 26, no. 3, 1989.

Index

Picture Credits

Photos supplied by Research Plus, Inc., Mill Valley, California

Cover photo by FPG, International
American Cancer Society, 20
AP/Wide World Photos, 8, 14, 15, 28, 29, 32, 48, 56, 61, 68
Sebastiao Barbosa, 63
Bettmann Archive, 54
Biological Photo Service, 16
British Airways, 40
George Dritsas/Light Images, 58
Helena Frost/FPG, 12
D.A. Glawe/Biological Photo Service, 27
Greenpeace/Plowden, 31
Mike J. Howell/Light Images, 34
Jim MacKenzie/World Resources Institute, 24
Montes De Oca, 6
National Aeronautics and Space Administration, 41, 50
National Oceanic and Atmospheric Administration, 10
Bill Pogue/Light Images, 25 (bottom), 30
Port Authority of NY & NJ, 60
Research Plus, Inc., 43
Reuters/Bettmann, 21
United Nations/Canadian Govt. Travel Bureau, 36
UPI/Bettmann, 18, 22, 25 (top), 26, 35, 47, 64

About the Author

Don Nardo is an actor, makeup artist, film director, composer, and teacher, as well as a writer. As an actor, he has appeared in more than fifty stage productions, including several Shakespeare plays. He has also worked before or behind the camera in twenty films. Several of his musical compositions, including a young person's version of H.G. Wells's *The War of the Worlds*, have been played by regional orchestras. Mr. Nardo's writing credits include short stories, articles, textbooks, screenplays, and several teleplays, including an episode of ABC's "Spenser: For Hire." In addition, his screenplay *The Bet* won an award from the Massachusetts Artists Foundation. Mr. Nardo lives with his wife and son on Cape Cod in Massachusetts.